Jacob Westland

Outlines of Logic

Jacob Westland

Outlines of Logic

ISBN/EAN: 9783337312398

Printed in Europe, USA, Canada, Australia, Japan

Cover: Foto ©berggeist007 / pixelio.de

More available books at **www.hansebooks.com**

OUTLINES OF LOGIC.

BY

JACOB WESTLUND,

PROFESSOR OF MATHEMATICS IN BETHANY COLLEGE,

LINDSBORG, KANSAS.

TOPEKA, KANSAS:
CRANE & CO., PUBLISHERS.
1896.

TABLE OF CONTENTS.

INTRODUCTION.

DEFINITION AND SCOPE OF LOGIC.

                                      PAGE.
1. DEFINITION OF LOGIC.................................... 1
2. UTILITY OF LOGIC.... 2
3. OPERATIONS OF THE MIND 2

CHAPTER I.

FUNDAMENTAL LAWS OF THOUGHT.

1. DEFINITION OF LAW OF THOUGHT........................... 4
2. FUNDAMENTAL LAWS OF THOUGHT.......................... 4
    1. Law of Identity............................... 4
    2. Law of Contradiction.......................... 5
    3. Law of Excluded Middle 5
    4. Law of Sufficient Reason 5

CHAPTER II.

CONCEPTS.

1. DEFINITION OF CONCEPT 7
2. CLASSIFICATION OF CONCEPTS 7
3. CONTENT AND EXTENT 8
4. RELATION OF CONCEPTS:
    1. Compatible Concepts 10
    2. Incompatible Concepts......................... 10
    3. Subordinate Concepts.......................... 11
    4. Coördinate Concepts........................... 12

CHAPTER III.

JUDGMENTS.

1. DEFINITION OF JUDGMENT 14
2. TERMS 14
3. CLASSIFICATION OF JUDGMENTS:
    1. Quality 15
    2. Quantity 15
    3. Relation 17
    4. Modality 19
4. DISTRIBUTION OF TERMS 20
5. IMMEDIATE INFERENCE:
    1. Synonymous Judgments 23
    2. Subalterns 24
    3. Opposition 25
    4. Conversion 29
6. SIMPLE AND COMPLEX PROPOSITIONS 33

CHAPTER IV.

SYLLOGISMS.

1. DEFINITION OF SYLLOGISM 35
2. CLASSIFICATION OF SYLLOGISMS 36
3. CATEGORICAL SYLLOGISMS:
    1. Definition 37
    2. Rules 37
    3. Explanation of Rules 38
    4. Figures 42
    5. Moods 43
4. HYPOTHETICAL SYLLOGISMS:
    1. Definition 49
    2. Moods 50
    3. Fallacies 52
    4. Reduction of Hypothetical to Categorical Syllogisms 53
5. DISJUNCTIVE SYLLOGISMS:
    1. Definition 54
    2. Rules 54
    3. Moods 54

	PAGE.
6. DILEMMA	55
7. COMPOUND SYLLOGISMS	56
8. ABRIDGED SYLLOGISMS:	
1. Enthymeme	58
2. Epichirema	59
3. Sorites	59

CHAPTER V.

FALLACIES.

1. DEFINITION OF FALLACY	63
2. CLASSIFICATION OF FALLACIES	63
3. LOGICAL FALLACIES:	
1. Fallacy of Equivocation	65
2. Fallacy of Composition	66
3. Fallacy of Division	66
4. Fallacy of Accident	67
5. Converse Fallacy of Accident	67
6. Fallacy of Many Questions	67
7. Fallacy of Amphibology	68
8. Fallacy of Positive and Negative Intention	68
4. MATERIAL FALLACIES:	
1. Begging the Question	69
2. Fallacy of False Cause	69
3. Fallacy of Irrelevant Conclusion	70
5. PARALOGISMS AND SOPHISMS	70

CHAPTER VI.

METHOD.

1. SCIENCE:	
1. Definition of Science	72
2. Requisites of a Science	72
3. Axioms	73
2. DEDUCTION AND INDUCTION:	
1. Definition of Method	73
2. Deduction	74
3. Induction	74

3. **DEFINITION:**
   1. Definition defined 76
   2. Rules for Definition 77
   3. Nominal and Real Definitions 80
   4. Description 80
4. **DIVISION:**
   1. Division defined 81
   2. Dichotomy 81
   3. Rules for Division 82
   4. Partition 84
5. **DEMONSTRATION:**
   1. Demonstration defined 85
   2. Rules for Demonstration 85
   3. Classification of Demonstrations 86
6. **ANALOGY** 90
7. **HYPOTHESIS** 91
8. **CLASSIFICATION OF SCIENCES** 91

EXERCISES 92–102

INTRODUCTION.

DEFINITION AND SCOPE OF LOGIC.

1. Definition of Logic.

Logic is the science of the formal laws of human thought.

Logic is the science which has for its object to investigate the laws of human thought apart from the other acts of the mind. It explains the laws and principles by which all reasoning must be governed. In all sciences the reasoning must be in accordance with the principles of logic, and although the method may be different in different sciences it must always conform to the laws of thought.

Logic is mainly a formal science, having for its object to ascertain and describe all the general forms in which thought presents itself without regard to any subject-matter. Logic differs from psychology in having for its object only the investigation of the formal laws of thought, while psychology treats of all the facts of the human mind and the laws by which its operations are guided.

2. Utility of Logic.

Logic does not teach us to think, but teaches us the laws by which our reasoning must be guided. All persons learn to think and to reason even before they know the name of logic, and thus unconsciously apply the principles of logic; but many questions are of so complex and difficult a nature that it is only by the aid of logic that we are able to detect what is correct or fallacious in the argument. The chief utility of logic thus consists in giving an invariable test of the correctness of an argument.

3. Operations of the Mind.

In approaching an argument the mind passes through the following intellectual processes: *Perception, abstraction, generalization, judgment,* and *reasoning.*

1. *Perception is the act of the mind by which it gains knowledge of external objects through the senses.*

The products of perception are called *percepts.* Thus my idea of my house, or of Boston, or of any particular object, is a percept.

2. *Abstraction is the act of the mind by which it draws a quality away from an object and considers it apart from the other peculiarities of the object.*

Thus the observing of the color of a certain object and making that a distinct object of thought to the exclusion of all the other qualities of the same object, is a process of abstraction.

3. *Generalization is the act of the mind by which it considers the qualities which are common to all the individuals of a group of objects and unites them into a single notion comprehending them.*

Thus if we consider the properties common to all kinds of triangles, disregarding difference in size or shape, the process is generalization.

A *concept* or *general notion* is the product of abstraction and generalization.

The concept *plant*, for instance, is formed by fixing our attention upon the properties common to all individual plants and disregarding all the points in which they differ. The concept *plant* thus embraces all individual plants, and is a name that may be applied to any one of them.

A concept is always *general*, a percept *particular*.

4. *Judgment is the act of the mind by which we compare two objects of thought, asserting whether they agree or not.*

The product of this operation is called a *judgment*. A judgment expressed in words is called a *proposition*.

5. *Reasoning is the act of the mind which consists in drawing conclusions from two or more judgments.*

An act of reasoning in its simplest logical form is called a *syllogism*.

CHAPTER I.

FUNDAMENTAL LAWS OF THOUGHT.

1. Definition of Law of Thought.

A law of thought is a necessary and universal principle by which all thought must be governed.

2. Fundamental Laws of Thought.

There are four fundamental laws of thought, on which all reasoning must ultimately depend. These laws are:

1. **The Law of Identity** (Principium identitatis).

2. **The Law of Contradiction** (Principium contradictionis).

3. **The Law of Excluded Middle** (Principium exclusi tertii).

4. **The Law of Sufficient Reason** (Principium rationis sufficientis).

1. THE LAW OF IDENTITY.— *Whatever is, is.*

This law may be expressed by the formula $A = A$. Its meaning is that *everything is identical with itself.* All the attributes of a thing must be consistent with each other and with the thing itself. In the proposition *All*

Americans are rational beings, the identity of *All Americans* with *some rational beings* is set forth.

2. THE LAW OF CONTRADICTION.—*Nothing can both be and not be.*

This law may be expressed by the formula $A \text{ not} = \text{not}—A$. The attributes of an object must not be inconsistent with each other nor with the thing itself. In the proposition, *No animals are plants*, we assert that *animals* are inconsistent with *plants*. A triangle may be either right-angled or not right-angled, but we cannot conceive that it should be both at the same time. If we say that a triangle is round, we evidently violate this law, because roundness is a quality inconsistent with a triangle.

3. THE LAW OF EXCLUDED MIDDLE.—*Everything must either be or not be.*

This law may be expressed by the formula, *A is either B or not—B.* It is impossible to conceive of any thing and any quality without affirming that the quality either belongs to the thing or does not belong to it. Gold, for instance, must be either a metal or not a metal. There is no third.

4. THE LAW OF SUFFICIENT REASON.—*For every consequent there must be a sufficient reason.*

If two propositions are connected in such a manner that the truth of one necessarily implies the truth of the other, the former is called the *reason* and the latter the *consequent*.

This law may be expressed by the formula, *If A is, B is.* Its meaning is, that for every proposition that is not intuitively true a sufficient reason must be given.

For instance, *If two triangles have equal bases and equal altitudes, they are equivalent.* Here the equivalence of the triangles is the consequent, and the reason why they are equivalent is that they have equal bases and equal altitudes.

CHAPTER II.

CONCEPTS.

1. Definition of Concept.

A concept or general notion is the consciousness in our mind of the attributes common to all the individuals of a certain group of objects.

Concepts are formed by abstraction and generalization, as has already been mentioned. As examples of concepts we may give the following: *Man, animal, book, triangle, plant, planet, heavenly body*, and *dog*.

2. Classification of Concepts.

Concepts may be divided into 1. *Positive* and *negative;* 2. *Absolute* and *Relative;* 3. *Concrete* and *Abstract*.

1. *a)* A *positive* concept is one in which the existence of a quality is asserted.

b) A *negative* concept is one in which the absence of a quality is asserted.

Thus, *organic* and *right-angled* are positive and *inorganic* and *not right-angled* negative concepts.

2. *a)* An *absolute* concept is one that can be thought of without reference to some other concept.

b) A *relative* concept is one that cannot be thought of without reference to some other concept.

Thus, *father*, *mother*, *son*, and *daughter* are relative concepts. We cannot think of father or mother without reference to a child, nor of son or daughter without reference to father or mother. *Metal*, *water*, and *triangle*, on the other hand, are terms which have no apparent relation to any other things, and which therefore are absolute.

3. *a)* A *concrete* concept is a name that can be applied to a thing.

b) An *abstract* concept is the name of a quality that belongs to a thing.

Thus *circle*, *table*, and *brick-house* are concrete; but *redness*, *hardness*, and *usefulness* abstract concepts.

3. Content and Extent.

Every concept has *content* and *extent*.

By the content of a concept is meant all the marks or attributes of the concept.

By the extent of a concept is meant all the individuals or objects it embraces.

Let us take the concept *insect*. The content of *insect*

consists of all the attributes which are necessarily possessed by all insects and by which an insect is distinguished from all other beings. By the *extent* of insect we mean all the different kinds of insects that exist.

When we compare two concepts that are related to one another, we observe that the concept which is poorer in content has the greater extent, and that the one that has the greater content has the smaller extent; or as it is usually expressed:

The content and extent of two concepts are in inverse ratio to each other.

In order to make this clear let us compare the two terms *fish* and *vertebrate*. The term vertebrate includes not only all the animals that are included under the term *fish*, but also *reptiles, birds, mammals*, etc. Consequently *vertebrate* has a greater extent than *fish*. On the other hand, all the properties that belong to vertebrates must necessarily belong to all fishes, and in addition to these there are many properties that belong exclusively to fishes and by which fishes are distinguished from all other vertebrates. Therefore *fish*, having a greater number of marks or attributes than *vertebrate*, has the greater content. *Vertebrate* is a term that may be applied to all fishes, and *fish* is an individual case of *vertebrate*. As another example let us take the two terms *plane figure* and *circle*. Of these the former has obviously the greater extent and the latter the greater content.

If two concepts are so related that one includes the other, as the concepts *vertebrate* and *fish*, the one that includes the other is called the *higher* concept, and the one that is included in the other is called the *lower* concept. Thus *vertebrate* is the *higher* and *fish* the *lower* concept.

4. Relation of Concepts.

Two or more concepts may be compared: 1st, with respect to content; and 2d, with respect to extent. In the first case they may be either *compatible* or *incompatible*. In the second case they may be either *subordinate* or *co-ordinate*.

1. COMPATIBLE CONCEPTS.—*Two concepts are said to be compatible when they both can be affirmed of the same subject, or both are included in the content of the same concept.*

The two terms *equilateral* and *right-angled*, for instance, may both be affirmed of a square, and are consequently compatible. *Large* and *heavy* are also two compatible terms, because they may be affirmed of the same subject.

2. INCOMPATIBLE CONCEPTS.—*Two concepts are said to be incompatible when they cannot both be affirmed of the same subject, or are not included in the content of the same concept.*

Incompatible concepts are either *contradictory* or *contrary*.

a) *Two concepts are contradictory when one is the negative of the other.*

For instance, *cold* and *not-cold*, *figure* and *not-figure*, *organic* and *inorganic*, etc. From a logical point of view, it is immaterial which one of two contradictory terms is considered positive and which negative. Each is the negative of the other.

b) *Two concepts are contrary when one not only implies a negation of the other, but also expresses some positive attribute.*

For instance, *man* and *woman*, *pentagonal* and *hexagonal*.

3. SUBORDINATE CONCEPTS.—*Of two concepts, one is said to be subordinate to the other when it is included in the extent of the other.*

If one concept is included in the extent of another, the former is called *species* and the latter *genus*. Thus, of the two terms *plant* and *tree*, *plant* is the genus, and *tree* is a species of the genus *plant*. The genus has always a greater extent than the species, *i. e.*, includes a greater number of individuals than the species. But as extent and content are in inverse ratio to each other, it follows that the species has greater content or a greater number of attributes than the genus. The species has not only all the attributes of the genus, but also other attributes by which it is distinguished from all other species of the same genus. The relation between two concepts of which one is subordinate to the other is shown by the diagram, where A (the outer circle) represents the genus and B (the inner circle) the species.

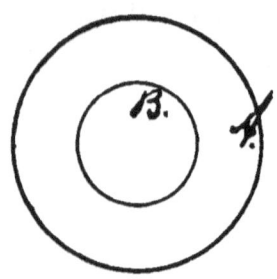

4. Coördinate Concepts. — *Two or more concepts are said to be co-ordinate to each other, when they are included in the extent of the same concept, but at the same time exclude each other.*

Thus, the two concepts *plant* and *animal* are coördinate to each other, being both species of the same genus *organic being*, and also excluding each other. The terms *plant* and *tree* are not coördinate. They are both included in the extent of *organic being*, but they do not exclude each other. As another example of coördinate terms we may take *fish*, *bird*, and *mammal*, all three being species of the genus *vertebrate*. The relation between coördinate terms is shown by the diagram, where A represents the genus and B, C, and D three of its species.

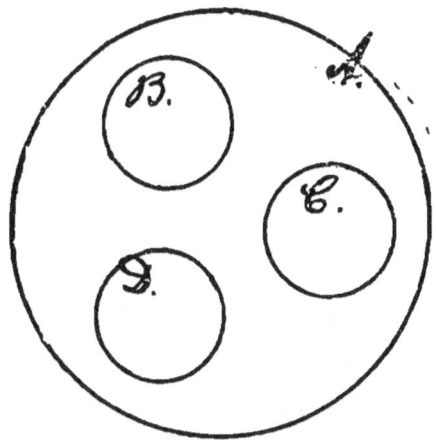

If one concept is subordinate to another, they must be compatible; and if two or more concepts are co-ordinate to each other, they must be incompatible.

The truth of this may be verified by taking the terms *parallelogram* and *quadrilateral* and the terms *plant* and *animal*. Of the two terms *parallelogram* and *quadrilateral*, the former is subordinate to the latter. But as all rectangles are parallelograms, and also all rectangles are quadrilaterals, we see that the two terms parallelogram and quadrilateral may both be affirmed of the same subject, *rectangle*. Hence *parallelogram* and *quadrilateral* are compatible.

The two terms *plant* and *animal* are evidently coördinate to each other, both being species of the genus *organic being*, and at the same time excluding each other. But we cannot find any subject of which they may both be affirmed. Hence they are incompatible.

CHAPTER III.

JUDGMENTS.

1. Definition of Judgment.

Judgment is that act of thought by which we compare two objects of thought, asserting whether they agree or not.

The product of this operation is called a *judgment*. A *logical proposition* is a judgment expressed in words. For instance, *All horses are mammals.*

2. Terms.

Every judgment contains two ideas, called the *terms* of the judgment. The term of which something is affirmed or denied is called the *subject*, and the term which is affirmed or denied of the subject is called the *predicate*. The word that expresses the connection between the subject and the predicate is called the *copula*.

Thus, in the judgment

Man is mortal,

man is the subject, *mortal* the predicate, and *is* the copula.

Of the two terms of a judgment the predicate is usually

a concept, and the subject may be either a concept or a percept. Thus, in the proposition *Insects are animals*, both the subject and the predicate are concepts. In the proposition *Chicago is a city*, the subject is a percept and the predicate a concept. Sometimes both terms may be percepts, as in *Chicago is not London*.

3. Classification of Judgments.

Judgments are classified according to *quality*, *quantity*, *relation*, and *modality*.

1. QUALITY.—According to *quality* judgments are divided into *affirmative* and *negative*.

a) An affirmative judgment is one in which the predicate is affirmed of the subject.

For instance,
    All horses are animals.

b) A negative judgment is one in which the predicate is denied of the subject.

For instance,
    No roses are animals.

2. QUANTITY.— According to *quantity* judgments are divided into *universal* and *particular*.

a) A universal judgment is one in which the predicate is affirmed or denied of the subject in its whole extent.

For instance,
    All men are mortal.

b) *A particular judgment is one in which the predicate is affirmed or denied of the subject only in part of its extent.*

For instance,

        Some animals are insects.

        No triangles are circles.

A judgment which has for its subject a singular term is sometimes called a *singular* judgment, as *Alexander was a conqueror.* All singular judgments, however, are universal, since in such a judgment the predicate is evidently affirmed or denied of the whole of the subject.

A proposition is said to be *indefinite* when it has no mark of quantity whatever, leaving it ambiguous whether it is universal or particular. In all such cases, however, the proper mark of quantity can be prefixed. Thus, the indefinite proposition *Man is mortal* means *All men are mortal.*

The combination of difference in quality with difference in quantity gives rise to four classes of judgments:

        Universal affirmative. A.
        Universal negative. E.
        Particular affirmative. I.
        Particular negative. O.

These four classes of judgments are designated by the letters *A*, *E*, *I*, and *O*. It is easy to remember what kind of judgment each letter represents by observing that A and I are the first two vowels of the Latin word *affirmo*, and E and O the vowels of *nego*. We give the following examples:

>All insects are animals. *A.*
>No men are gods. *E.*
>Some men are wise. *I.*
>Some men are not wise. *O.*

In passing from a particular affirmative to a particular negative judgment, we prefix *not* to the predicate. When we pass from a universal affirmative to a universal negative judgment, however, this is not sufficient. In that case the negative adjective *no* must be prefixed to the subject. Let us take the universal affirmative judgment *All men are rational.* By prefixing *not* to the predicate we have *All men are not rational*, which may be particular and may imply that some men may be rational. It is therefore not a complete negation of the universal affirmative judgment *All men are not rational.* Hence, in order to express a complete denial of the universal affirmative judgment we must prefix *no* to the subject. Thus, *No men are rational.*

3. RELATION.—According to *relation* judgments are divided into *categorical, hypothetical,* and *disjunctive.*

a) *A categorical judgment is one in which the predicate is unconditionally affirmed or denied of the subject.*

The simplest form of a categorical judgment is,

>*S is P.*

For instance,

All trees are plants.

Some heavenly bodies are not planets.

b) A hypothetical judgment is one in which the predicate is affirmed or denied of the subject conditionally.

The simplest form of a hypothetical judgment is,

If A is B, C is D.

A hypothetical judgment thus consists of two categorical judgments connected by the conjunction *if*. The first, or the one that expresses the condition, is called the *antecedent*, and the other the *consequent*.

For instance,

If rain does not come, the crops will fail.

Here, *If rain does not come* is the antecedent, and *the crops will fail* is the consequent.

A hypothetical judgment can always be changed to a categorical judgment of exactly the same meaning, having for its subject the antecedent and for its predicate the consequent of the hypothetical judgment.

Thus, the hypothetical judgment

If a triangle is equilateral, it is equiangular

can be converted into the categorical judgment

All equilateral triangles are equiangular.

c) A disjunctive judgment is one that expresses an alternative.

The simplest form of a disjunctive judgment is
S is either P or not P.

The disjunctive judgment has instead of a single predicate two alternatives or more, of which one must be asserted of the subject to the exclusion of any other alternative. For instance,

John is either in the house or not in the house.

This triangle is either right-angled, obtuse-angled, or acute-angled.

A disjunctive judgment is called *divisive* when the predicate expresses all the species of the subject. For instance,

Organic beings are divided into animals and plants.
Triangles are divided into right-angled and oblique-angled.

The divisive judgment is disjunctive only in form, but categorical in sense. It is, in reality, composed of two or more particular judgments. Thus, the judgment *Triangles are divided into right-angled and oblique-angled* is composed of the two particular judgments

Some triangles are right-angled.
Some triangles are oblique-angled.

4. MODALITY.—According to *modality*, or the degree of certainty, judgments are divided into *apodictic, problematic*, and *assertory*.

a) An apodictic judgment is one which expresses the

combination between the subject and the predicate as a necessity.

$$S \text{ must be } P.$$

For instance,

An equilateral triangle must be equiangular.

b) A problematic judgment is one which expresses the combination between the subject and the predicate as a possibility.

$$S \text{ may be } P.$$

For instance,

Mars may be inhabited.

c) An assertory judgment is one which expresses the combination between the subject and the predicate as a fact to be taken for granted.

$$S \text{ is } P.$$

For instance,

This dog is mad.

4. Distribution of Terms.

A term is said to be distributed when it is taken universally or in its whole extent.

For instance, in the judgment *All animals are organic beings*, the term *animal* is taken universally or in its whole extent, and is therefore distributed.

1. With regard to the *subject* we have the following rules:

a) In   All S are P   (A)
    and No S are P   (E)

the subject is *distributed*, both judgments being universal.

b) In   *Some S are P*   (*I*)
and   *Some S are not P*   (*O*)

the subject is *not distributed*, both judgments being particular.

2. With regard to the *predicate* we have the following rules:

a) In   *All S are P*   (*A*)

the predicate is *not distributed*. It is evident that the whole of P is not considered, as P may contain many other things besides S.

b) In   *Some S are P*   (*I*)

the predicate is *not distributed*, as is shown by the same reasoning as for A.

c) In   *No S are P*   (*E*)

the predicate is *distributed*. In order to assert that no part of S belongs to any part of P, it is evident that the whole of P must be considered.

d) In   *Some S are not P*   (*O*)

the predicate is *distributed*. The same reasoning applies here as for E. We must consider the whole of P in order to assert that no part of it belongs to *some P* in question.

REMARKS.

1. A distributes the predicate in case the subject and the predicate are co-extensive, *i. e.*, have exactly the same extent.

For instance,

All equilateral triangles are equiangular.

2. *I* distributes the predicate in case the subject is the genus and the predicate one of its species. For instance, Some animals are vertebrates.

For the distribution of terms in the four categorical judgments we have then the following rules:

1. *Universal judgments distribute the subject; particular judgments do not.*
2. *Negative judgments distribute the predicate; affirmative judgments do not.*

These rules may be stated by the following schedule:

SUBJECT.	PREDICATE.
A. Distributed.	*Undistributed.*
E. Distributed.	*Distributed.*
I. Undistributed.	*Undistributed.*
O. Undistributed.	*Distributed.*

In the diagrams given below the distribution of the subject and the predicate in the four categorical judgments is shown. S represents the subject and P the predicate.

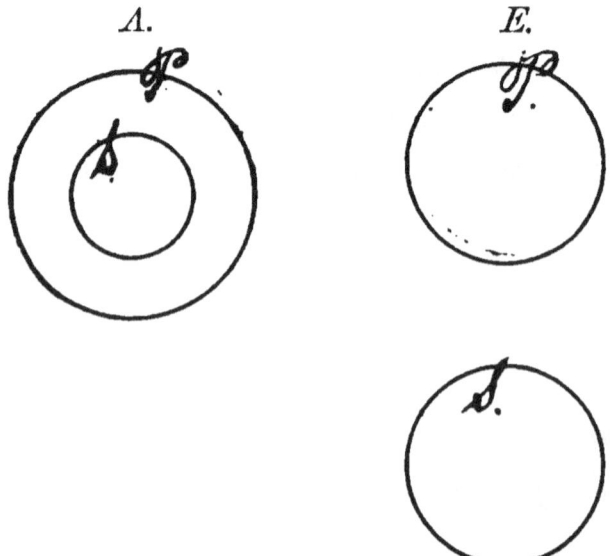

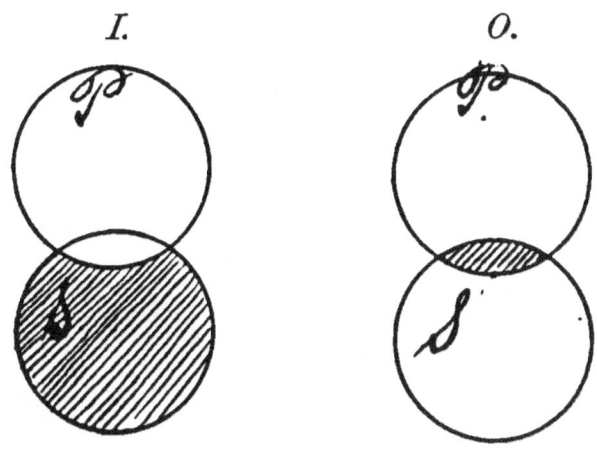

5. Immediate Inference.

Immediate inference is that act of thought by which we transform one judgment into another and from the validity or invalidity of one infer the validity or invalidity of the other.

We will treat immediate inference under the following heads:

1. *Synonymous Judgments.*
2. *Subalterns.*
3. *Opposition.*
4. *Conversion.*

1. SYNONYMOUS JUDGMENTS.—*Two judgments are synonymous when they express the same fact in different words.*

The wording of a proposition may evidently be changed in many different ways so as to give a new proposition, differing only in form but not in sense from the given one. We may, for instance, substitute for either the subject or the predicate equivalent terms; or change from a categorical to a hypothetical proposition, and conversely; or instead of affirming one thing, deny its opposite. Evidently both are true or both false at the same time.

For instance:

 (True) This is a triangle.
 (True) This is a figure having three sides.

 (False) All triangles are equilateral.
 (False) No triangles are not equilateral.

 (True) Damp gunpowder will not explode.
 (True) If gunpowder is damp, it will not explode.

2. SUBALTERNS.—*Two judgments are said to be subalterns when they have the same subject, the same predicate, and the same quality, but one is universal and the other particular.*

Thus, *A* and *I* are a pair of subalterns; also *E* and *O*. *I* and *O* are called the *subalternates* of *A* and *E* respectively, each of which is a *subalternans*.

From the truth of the universal we infer the truth of the particular, and from the falsity of the particular we infer the falsity of the universal. But the truth of the particular does not always include the truth of the universal; nor does the falsity of the universal always include the falsity of the particular.

For instance,
> (True) All men are mortal. (*A*)
> (True) Some men are mortal. (*I*)
> (True) No animal is rational. (*E*)
> (True) Some animals are not rational. (*O*)
>
> (False) Some plants are animals. (*I*)
> (False) All plants are animals. (*A*)
>
> (False) Some triangles are not figures. (*O*)
> (False) No triangles are figures. (*E*)

But the truth of the particular judgment
> Some animals are insects (*I*)

does not involve the truth of the universal
> All animals are insects. (*A*)

Nor can we from the falsity of the universal judgment
> No figures are triangles (*E*)

infer the falsity of the particular
> Some figures are not triangles. (*O*)

Hence we conclude from the truth of *A* and *E* to the truth of *I* and *O* respectively, and from the falsity of *I* and *O* to the falsity of *A* and *E* respectively; but not from the falsity of *A* and *E* to the falsity of *I* and *O* respectively, nor from the truth of *I* and *O* to the truth of *A* and *E* respectively.

3. OPPOSITION. — *Opposition takes place between two judgments when they have the same subject and the same predicate, but opposite quality.*

There are three kinds of opposition depending on the quantity of the judgments, viz., *contrary*, *contradictory*, and *subcontrary*.

a) Contrary.—If both judgments are universal, the opposition is said to be *contrary*, or the judgments are *contraries* each of the other.

Two contrary judgments cannot both be true, but they may both be false.

Hence the truth of one involves the falsity of the other, but the falsity of one does not necessarily involve the truth of the other.

For instance,

(True) All trees are plants. (A)

(False) No trees are plants. (E)

(True) No animals are plants. (E)

(False) All animals are plants. (A)

But from the falsity of

All animals are insects (A)

we cannot infer the truth of

No animals are insects. (E)

Nor can we from the falsity of

No animals are birds (E)

infer the truth of

All animals are birds. (A)

Hence we conclude from the truth of A to the falsity of E and from the truth of E to the falsity of A, but not from the falsity of A to the truth of E, nor from the falsity of E to the truth of A.

b) Contradictory.— If one judgment is universal and the other particular, the opposition is said to be *contradictory*, or the judgments are *contradictories* each of the other.

Of two contradictory judgments one must be true and the other false.

Hence the truth of one involves the falsity of the other, and the falsity of one involves the truth of the other.

For instance,
    (True) All plants are organic beings. (A)
    (False) Some plants are not organic beings. (O)
    (True) No triangles are squares. (E)
    (False) Some triangles are squares. (I)
    (True) Some animals are not birds. (O)
    (False) All animals are birds. (A)
    (True) Some plants are water-plants. (I)
    (False) No plants are water-plants. (E)

Hence we conclude from the truth or falsity of A, E, I, and O to the falsity or truth of O, I, E, and A respectively.

c) Subcontrary.— If both judgments are particular, the opposition is said to be *subcontrary*, or the judgments are *subcontraries* each of the other.

Two subcontrary judgments may both be true, but they cannot both be false.

Hence from the falsity of one we infer the truth of the other, but the truth of one does not necessarily involve the falsity of the other.

For instance,

      (False) Some triangles are not figures. (*O*)
      (True) Some triangles are figures. (*I*)

      (False) Some plants are animals. (*I*)
      (True) Some plants are not animals. (*O*)

But from the truth of

      Some heavenly bodies are planets (*I*)

we cannot infer the falsity of

      Some heavenly bodies are not planets. (*O*)

Nor can we from the truth of

      Some animals are not fishes (*O*)

infer the falsity of

      Some animals are fishes. (*I*)

Hence we conclude from the falsity of *O* and *I* to the truth of *I* and *O* respectively, but not from the truth of *O* and *I* to the falsity of *I* and *O* respectively.

The relations between the four judgments *A*, *E*, *I*, and *O* are shown by the following schedule:

4. CONVERSION.—*A judgment is said to undergo conversion or to be converted when its subject and predicate are interchanged.*

If the given judgment is true, the new judgment must also be true.

There are three kinds of conversion: *simple conversion, conversion by limitation,* and *conversion by contraposition.*

a) *Simple conversion.*—A judgment is simply *converted* when its subject and predicate are interchanged, the quality and quantity remaining the same.

For instance,

  No metals are compounds. (*E*)
No compounds are metals. (*E*)

 Some flowers are yellow. (*I*)
Some yellow things are flowers. (*I*)

But from the judgment

 All metals are elements (*A*)
we cannot infer that
All elements are metals.

Nor can we pass from

 Some plants are not water-plants (*O*)
to
Some water-plants are not plants.

Hence only *universal negative* and *particular affirmative* judgments can be simply converted.

b) Conversion by limitation.—A judgment is said to be *converted by limitation* when its subject and predicate are interchanged, the quality remaining the same, but the quantity being changed.

For instance,

All men are mortal. (*A*)
Some mortal beings are men. (*I*)

But from the judgment

Some animals are not insects (*O*)
we cannot pass to
No insects are animals.

Hence all *universal affirmative* judgments can be converted by limitation. To *particular negative* judgments neither simple conversion nor conversion by limitation can be applied.

There are, however, some universal affirmative judgments that can be simply converted; namely, all those in which the subject and the predicate are co-extensive. To that class belong all logical definitions.

For instance,

A quadrilateral is a figure having four sides.
All figures having four sides are quadrilaterals.

All equilateral triangles are equiangular.
All equiangular triangles are equilateral.

c) *Conversion by contraposition.*—We are said to convert a judgment by *contraposition* when we first change the quality and for the predicate substitute its contradictory and then apply simple conversion.

By the first process we pass from the affirmation of one thing to the denial of its opposite. For instance,

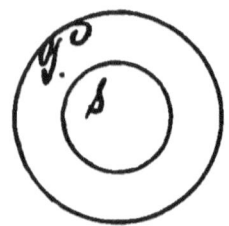
All metals are elements (*A*)
No metals are not-elements (*E*)

and then by simple conversion
No not-elements are metals (*E*)

Some animals are not insects (*O*)
Some animals are not-insects (*I*)

and by simple conversion
Some not-insects are animals (*I*).

In the particular negative judgment we thus simply transfer the negative particle from the copula to the predicate and then apply simple conversion.

Hence all *universal affirmative* and *particular negative* judgments can be converted by contraposition.

A similar process may be applied to the universal negative judgment, though in that case we can only convert by limitation. For instance,

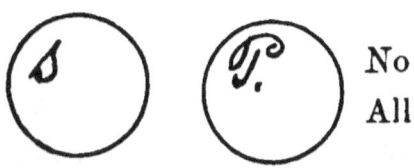
No fishes are birds. (*E*)
All fishes are not-birds. (*A*)

and by conversion by limitation

Some not-birds are fishes. (*I*)

For conversion we have then the following rules:

I. *Only universal negative and particular affirmative judgments can be simply converted.*

II. *All universal affirmative judgments can be converted by limitation.*

III. *Particular negative judgments can only be converted by contraposition.*

6. Simple and Complex Propositions.

A *simple* proposition is one that has only one subject and one predicate. For instance,

Gold is a metal.

A *complex* proposition is one that has more than one

subject, or more than one predicate, or both. For instance,

    Birds and fishes are animals.

In this example there are evidently two categorical propositions combined in one, viz.,

    Birds are animals

and   Fishes are animals.

The only complex propositions with which logic is directly concerned are the hypothetical and the disjunctive propositions, which have already been described.

CHAPTER IV.

SYLLOGISMS.

1. Definition of Syllogism.

Syllogism is the process by which two objects of thought are compared through their relation to a third.

Every syllogism contains three terms, the *major term*, the *middle term*, and the *minor term*. The relation between the three terms is expressed by three judgments, of which two are called the *premises* and the third the *conclusion*. In one premise the middle term is compared with the major term, in the other premise it is compared with the minor term, and in the conclusion the major and minor terms are compared. The premise containing the major term is called the *major premise*, and the premise containing the minor term is called the *minor premise*. The middle term, being only the medium of comparison between the two other terms, occurs only in the premises, but not in the conclusion. The minor term is always the *subject* of the conclusion, and the major term is always the *predicate* of the conclusion.

The *minor* and *major* terms are so called because the major term has usually greater extent than the minor term. The three terms of a syllogism are usually represented by the letters *P*, *M*, and *S*. *P* designates the major term, being the *predicate* of the conclusion; *M* denotes the middle term; and *S* denotes the minor term, being the *subject* of the conclusion.

The three judgments of a syllogism are usually arranged in the following order:

Major premise. All men are rational.
Minor premise. All Americans are men.
Conclusion. All Americans are rational.

In the example given above, *men* is the middle term, *rational* the major term, and *Americans* the minor term.

The syllogism may also be defined as the act of thought by which from two given judgments, called the premises, we draw or infer a third judgment, called the conclusion. Syllogism is also called *mediate inference*, and differs from *immediate inference*, described in the preceding chapter, mediate inference being made through a medium or a middle term.

2. Classification of Syllogisms.

Syllogisms are divided into *categorical*, *hypothetical*, and *disjunctive*.

1. *A categorical syllogism is a syllogism having for its major premise a categorical judgment.*

2. *A hypothetical syllogism is a syllogism having for its major premise a hypothetical judgment.*

3. *A disjunctive syllogism is a syllogism having for its major premise a disjunctive judgment.*

Examples:

Categorical. $\begin{cases} \text{M is P.} \\ \text{S is M.} \\ \therefore \text{S is P.} \end{cases}$

Hypothetical. $\begin{cases} \text{If A is B, C is D.} \\ \text{A is B.} \\ \therefore \text{C is D.} \end{cases}$

Disjunctive. $\begin{cases} \text{A is either B or not-B.} \\ \text{A is B.} \\ \therefore \text{A is not not-B.} \end{cases}$

3. Categorical Syllogisms.

1. DEFINITION.—*A categorical syllogism is a syllogism having for its major premise a categorical judgment.*

The minor premise and the conclusion are also categorical judgments.

2. RULES.—A general rule for the syllogism is an axiom known as the *dictum de omni et nullo* of Aristotle. This axiom may be stated thus:

Whatever is affirmed or denied of a whole class may also be affirmed or denied of any individual contained in that class.

The special rules of the categorical syllogism are:

I. *The syllogism must contain three and only three terms.*

II. *The syllogism must contain three and only three judgments.*

III. *The middle term must be distributed at least in one of the premises.*

IV. *In order that a term may be distributed in the conclusion it must be distributed in one of the premises.*

V. *From two negative premises no conclusion can be drawn.*

VI. *From two particular premises no conclusion can be drawn.*

VII. *If one premise is negative the conclusion will be negative.*

VIII. *If one premise is particular the conclusion will be particular.*

3. EXPLANATION OF THE RULES.—The first and second rules need no further explanation.

3d rule. If the middle term were not distributed in at least one of the premises, it might happen that the minor and major terms are compared with different parts of the middle term, and therefore the middle term would no longer be a medium of comparison. For instance,

                 All P are M
                 All S are M.

Here the middle term is not distributed. P is one part

of M and S is another part of M, and these parts may or may not coincide. No relation can be established between S and P, as S may fall wholly without, or wholly within, or partly without and partly within P, as is seen in the diagram.

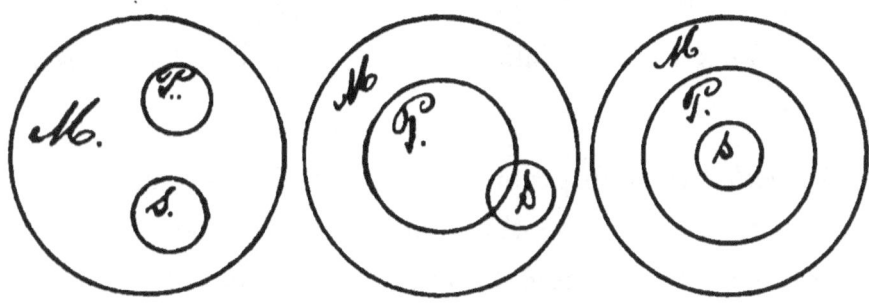

4th rule. If either the major or minor term is not distributed in the premise where it occurs, it must not be distributed in the conclusion. It is evident that we are only enabled to infer something about that part of either the major or minor term which has been compared with the middle term in the premise. In the syllogism

>All insects are animals
>No dogs are insects
>∴ No dogs are animals

the major term *animals* is not distributed in the major premise, but is distributed in the conclusion. This argument is consequently fallacious. This fallacy is called an *illicit process of the major term.*

Again, in the example

>All flies are insects
>All flies are animals
>∴ All animals are insects

the minor term *animals* is distributed in the conclusion, but not in the minor premise. Hence the argument is false. This kind of fallacy is called an *illicit process of the minor term.*

5th rule. If both premises are negative, no conclusion can be drawn, because the middle is no longer a medium of comparison between the minor and major terms. For instance, from the premises

>No M are P
>No S are M

no conclusion can be drawn as regards the relation between S and P, as S may fall wholly within, or wholly without, or partly within and partly without P, as is seen in the diagram.

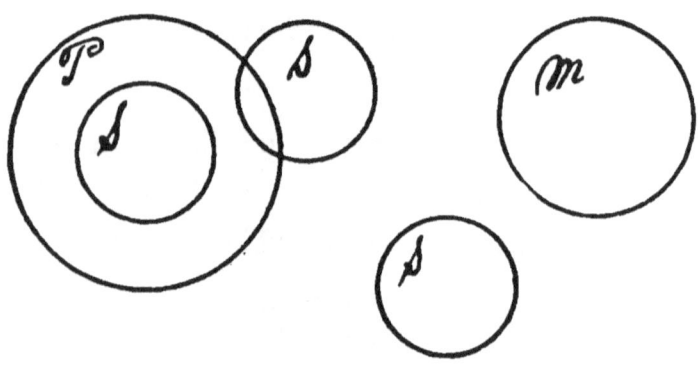

6th rule. If both premises are particular, no conclusion can be drawn, because no relation can be established between two terms that are only partly connected with a third. From the premises

　　　　Some M are P
　　　　Some S are M

no conclusion can be drawn; for, as is shown by the diagram, S may fall wholly without, or wholly within, or partly without and partly within P.

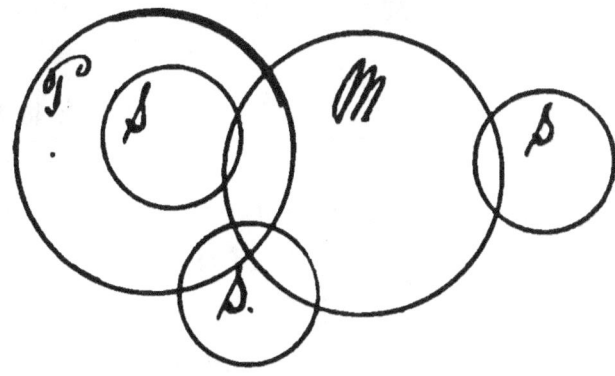

7th rule. If the minor premise, for instance, be negative, thus expressing a disagreement between the minor and middle terms, and the major premise affirmative, expressing an agreement between the major and middle terms, the conclusion must necessarily express a disagreement between the minor and major terms, *i. e.*, the conclusion must be negative. And in the same way if the major premise is negative and the minor premise affirmative.

8th rule. This rule is in fact a corollary of the third and fourth rules.

4. FIGURES.— The three terms of a syllogism may be arranged in different ways. It is evident that the middle term can have only four different positions, and hence there are four different ways, or, as they are called, *figures* in which the terms of a syllogism may be arranged. These four figures of the syllogism are shown in the following scheme:

1	2	3	4
M is P. S is M.	P is M. S is M.	M is P. M is S.	P is M. M is S.
S is P.	S is P.	S is P.	S is P.

In the *first* figure the middle term is the subject of the major premise and the predicate of the minor premise.

In the *second* figure the middle term is the predicate of both the major and the minor premises.

In the *third* figure the middle term is the subject of both the major and the minor premises.

In the *fourth* figure the middle term is the predicate of the major premise and the subject of the minor premise.

The first three figures were proposed by *Aristotle*, and hence they are usually called the *Aristotelian figures*. The fourth figure, proposed by *Galen*, is really an inversion of the first figure and is comparatively useless, because the same conclusions can be obtained more naturally by using the first figure.

5. MOODS.—As every syllogism must contain two premises and each premise may be either *universal affirmative*, *universal negative*, *particular affirmative*, or *particular negative*, there would be in each figure sixteen different forms of the syllogisms, or, as they are called, *moods*, depending on the quality and quantity of the premises. But the number of moods in each figure is limited by the rules of the syllogism mentioned above, and thus omitting all moods which violate these rules and all moods which are useless, being included in other moods, there will remain only *nineteen*. As an artificial aid in memorizing these nineteen possible moods the following mnemonic verses have been invented:

Fig. 1. *Barbara, Celarent, Darii, Ferioque*, prioris;

Fig. 2. *Cesare, Carnestres, Festino, Baroko*, secundæ;

Fig. 3. Tertia, *Darapti, Disamis, Datisi, Felapton, Bokardo, Ferison* habet; quarta insuper addit,

Fig. 4. *Bramantip, Camenes, Dimaris, Fesapo, Fresison*.

Each of the italicized names represents a mood, the vowels of each name standing for the three judgments of the syllogism. Thus for instance *Cesare* signifies the mood of the second figure, which has E for the major premise, A for the minor, and E for the conclusion.

First Figure.

This is the only figure in which the conclusion can be

universal affirmative. With regard to the premises the following rules will be observed:

a) *The major premise must be universal.*
b) *The minor premise must be affirmative.*

These special rules can easily be deduced from the general rules of the syllogism.

The four valid moods in this figure are:

Barbara, Celarent, Darii, and *Ferio.*

EXAMPLES.

Barbara.

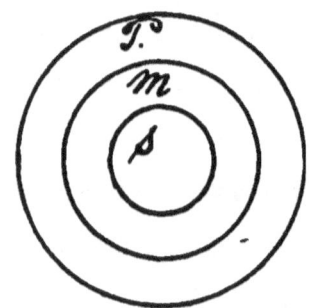

All men are mortal.
All Americans are men.
∴ All Americans are mortal.

Celarent.

No quadrilaterals are circles.
All parallelograms are quadrilaterals.
∴ No parallelograms are circles.

Darii.

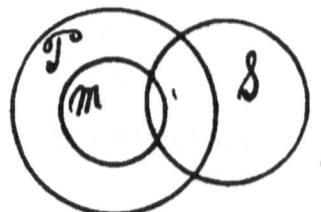

All mammals have red blood.
Some animals are mammals.
∴ Some animals have red blood.

Ferio.

No insects are warm-blooded.
Some animals are insects.
∴ Some animals are not warm-blooded.

Second figure.

In this figure the conclusion is always negative. For the premises we have the following special rules:

a) *The major premise must be universal.*
b) *One of the premises must be negative.*

The four valid moods of this figure are:

Cesare, Camestres, Festino, and *Baroko.*

EXAMPLES.

Cesare.

No trapezoid is equilateral.
All squares are equilateral.
∴ No squares are trapezoids.

Camestres.

All men are rational.
No apes are rational.
∴ No apes are men.

Festino.

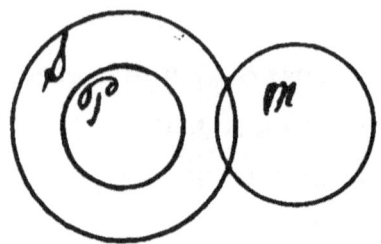

No planets are self-luminous.
Some heavenly bodies are self-luminous.
∴ Some heavenly bodies are not planets.

Baroko.

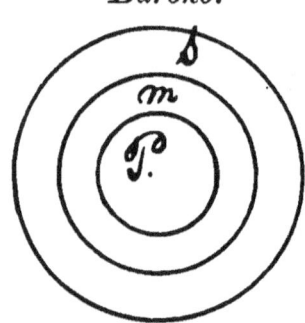

All horses are mammals.
Some animals are not mammals.
∴ Some animals are not horses.

Third Figure.

In this figure the conclusion is always particular. For the premises we have the following rule:

The minor premise must be affirmative.

Six moods are possible, viz.:

Darapti, Disamis, Datisi, Felapton, Bokardo, and *Ferison.*

EXAMPLES.

Darapti.

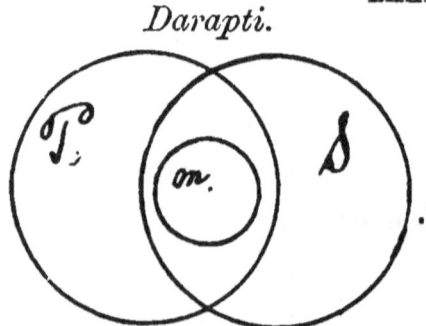

All whales are mammals.
All whales live in water.
∴ Some animals living in water are mammals.

Disamis.

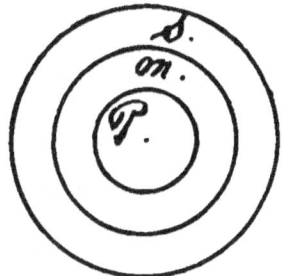

Some parallelograms are rectangles.
All parallelograms are quadrilaterals.
∴ Some quadrilaterals are rectangles.

Datisi.

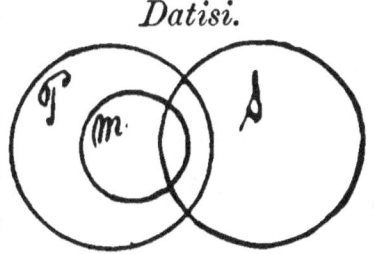

All parallelograms are quadrilaterals.
Some parallelograms are equilateral.
∴ Some equilateral figures are quadrilaterals.

Felapton.

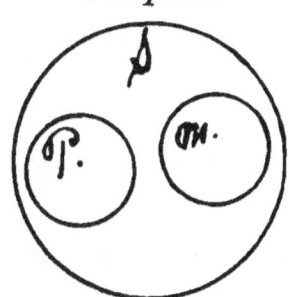

No man is omniscient.
All men are rational.
∴ Some rational beings are not omniscient.

Bokardo.

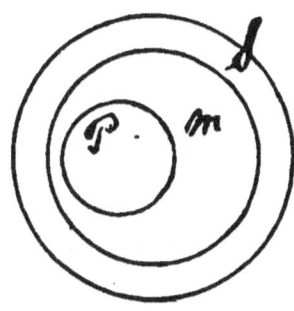

Some plants are not trees.
All plants are living beings.
∴ Some living beings are not trees.

Ferison.

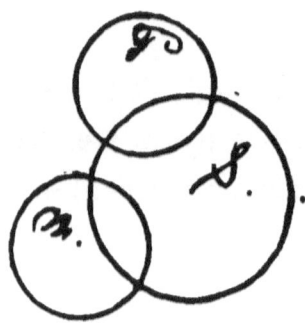

No animals are plants.
Some animals live in water.
∴ Some organisms living in water are not plants.

Fourth Figure.

In this figure there are five valid moods, viz.:
Bramantip, Camenes, Dimaris, Fesapo, and *Fresison.*

EXAMPLES.

Bramantip.

All fishes breathe by gills.
All animals breathing by gills are cold-blooded.
∴ Some cold-blooded animals are fishes.

Camenes.

All men are mortal.
No mortal being is omniscient.
∴ No omniscient being is a man.

Dimaris.

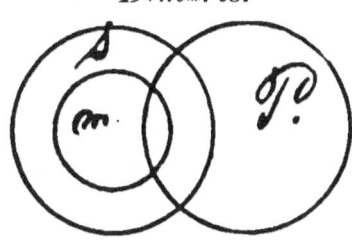

Some taxes are oppressive.
All oppressive things should be repealed.
∴ Some things which should be repealed are taxes.

Fesapo.

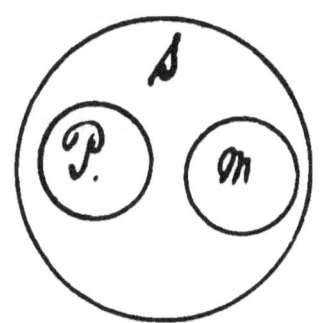

No immoral acts are proper amusements.
All proper amusements are designed to give pleasure.
∴ Some things designed to give pleasure are not immoral acts.

Fresison.

No birds have gills.
Some animals having gills are vertebrates.
∴ Some vertebrates are not birds.

4. Hypothetical Syllogisms.

1. DEFINITION.—*A hypothetical syllogism is a syllogism which has for its major premise a hypothetical judgment.*

The minor premise and the conclusion are usually categorical judgments. If all three judgments are hypothetical, the syllogism follows the same rules as the categorical syllogism, to which it can easily be reduced.

For instance,

>If a man violates the laws, he ought to be punished.
>If a man commits murder, he violates the laws.
>∴ If a man commits murder, he ought to be punished.

This can easily be put in the form of a categorical syllogism as follows:

>A man that violates the laws ought to be punished.
>A murderer violates the laws.
>∴ A murderer ought to be punished.

In the following we will therefore only consider hypothetical syllogisms in which the minor premise and the conclusion are categorical judgments.

2. MOODS.—Hypothetical syllogisms are divided into *constructive* and *destructive*, according as the minor premise is affirmative or negative. The first form is also called the *modus ponens*, or the mood that affirms, and the second the *modus tollens*, or the mood that denies.

a) Modus ponens.— For this mood we have the following rule:

If the antecedent be affirmed, the consequent must be affirmed. The minor premise affirms the antecedent and the conclusion affirms the consequent.

The general form of a constructive hypothetical syllogism is

If A is B, C is D.
A is B.
∴ *C is D.*

For instance,
  If a triangle is equilateral, it is equiangular.
  This triangle is equilateral.
∴ This triangle is equiangular.

  If he has a fever, he is sick.
  He has a fever.
∴ He is sick.

b) Modus tollens.— For this mood we have the following rule:

If the consequent be denied, the antecedent must be denied. The minor premise denies the consequent, and the conclusion denies the antecedent.

The general form of a destructive hypothetical syllogism is

If A is B, C is D.
C is not D.
∴ *A is not B.*

For instance,
  If a triangle is equilateral, it is equiangular.
  This triangle is not equiangular.
∴ This triangle is not equilateral.

If a man is a murderer, he ought to be punished.
This man ought not to be punished.
∴ This man is not a murderer.

3. FALLACIES.—If the minor premise either *affirms the consequent* or *denies the antecedent*, a fallacy of argument arises. If we affirm the consequent, we may not therefore affirm the antecedent, because the consequent might follow from some other antecedent as well as from the one given; or, as we might express it, a given effect may be produced by several different causes. For the same reason it is evident that we cannot pass from the denial of the antecedent to the denial of the consequent. Thus the argument,

If he has a fever, he is sick.
He is sick.
∴ He has a fever.

is fallacious. If a person is sick, it does not necessarily follow that he has a fever. He may be sick from some other cause. For the same reason the argument,

If he has a fever he is sick.
He has not a fever.
∴ He is not sick.

is fallacious.

There is one exception to this rule, and that is in case the given condition is the only condition of the consequent. In such a case we may pass from the affirmation of the consequent to the affirmation of the antecedent, or from the denial of the antecedent to the denial of the consequent. For instance,

>If a triangle is equilateral, it is equiangular.
>This triangle is equiangular.
>∴ This triangle is equilateral.

>If a triangle is equilateral, it is equiangular.
>This triangle is not equilateral.
>∴ This triangle is not equiangular.

In the above examples the two terms *equilateral triangle* and *equiangular triangle* are evidently co-extensive.

4. REDUCTION OF HYPOTHETICAL TO CATEGORICAL SYLLOGISMS.— As we have already seen, every hypothetical judgment can be converted into a universal affirmative judgment. Hence every hypothetical syllogism can be reduced to the categorical form and will consequently follow the rules laid down for the categorical syllogisms. In order to illustrate this we take the following example:

>If an animal is a mammal, it has red blood.
>All horses are mammals.
>∴ All animals have red blood.

By changing the major premise into a categorical judgment we obtain a categorical syllogism in the mood *Barbara*.

>All mammals have red blood.
>All horses are mammals.
>∴ All horses have red blood.

5. Disjunctive Syllogisms.

1. DEFINITION.—*A disjunctive syllogism is a syllogism which has for its major premise a disjunctive judgment.*

The minor premise and the conclusion are categorical judgments.

2. RULES.—The general rule governing all disjunctive syllogisms is:

If one or more alternatives be affirmed, the rest must be denied, and if one or more alternatives be denied, the rest must be affirmed.

This rule follows immediately from the *law of excluded middle*.

3. MOODS.—There are two moods, viz., *modus ponendo tollens* (the mood which by affirming denies) and *modus tollendo ponens* (the mood which by denying affirms), according as the minor premise is affirmative or negative.

a) *Modus ponendo tollens.*—The general form of this mood is

A is either B or not-B.
A is B.
∴ A is not not-B.

For instance,

A triangle is either right-angled, acute-angled, or obtuse-angled.
This triangle is right-angled.
∴ This triangle is neither acute-angled nor obtuse-angled.

b) *Modus tollendo ponens.*—The general form of this mood is

> *A is either B or not-B.*
> *A is not not-B.*
> ∴ *A is B.*

For instance,

> A triangle is either right-angled, acute-angled, or obtuse-angled.
>
> This triangle is neither right-angled nor acute-angled.
>
> ∴ This triangle is obtuse-angled.

6. Dilemma.

A dilemma is a syllogism having for its major premise a hypothetical judgment and for its minor premise a disjunctive judgment.

There are several different forms of the dilemma. We will only give one of the more common forms, in which the major premise is a hypothetical judgment whose consequent is disjunctive. This form of the dilemma may be stated thus:

> *If A is, either B or C is.*
> *Now neither B nor C is.*
> ∴ *A is not.*

This is in fact a destructive hypothetical syllogism. All

possible alternatives of the consequent are denied, therefore the antecedent must also be denied. For instance,

> If this triangle is not right-angled, it must be either obtuse-angled or acute-angled.
> Now it is neither obtuse-angled nor acute-angled.
> ∴ It must be right-angled.

7. Compound Syllogisms.

A series of syllogisms combined together in such a manner that the conclusion of the first is taken as a premise of the second and so on is called a compound syllogism or a poly-syllogism.

When the conclusion of one syllogism is used as a premise of another syllogism, the former syllogism is called a *pro-syllogism* and the latter an *epi-syllogism*. The conclusion of a pro-syllogism may be either the major or the minor premise of the epi-syllogism, as is seen by the following examples:

1.	2.
All C are D.	All B are C.
All B are C.	All A are B.
∴ All B are D.	∴ All A are C.
All B are D.	All C are D.
All A are B.	All A are C.
∴ All A are D.	∴ All A are D.

For A, B, C, and D let us take the terms *square, parallelogram, quadrilateral,* and *figure,* and we have the following compound syllogisms:

1.

All quadrilaterals are figures.
All parallelograms are quadrilaterals.
∴ All parallelograms are figures.

All parallelograms are figures.
All squares are parallelograms.
∴ All squares are figures.

2.

All parallelograms are quadrilaterals.
All squares are parallelograms.
∴ All squares are quadrilaterals.

All quadrilaterals are figures.
All squares are quadrilaterals.
∴ All squares are figures.

8. Abridged Syllogisms.

An abridged syllogism is a syllogism (either simple or compound) in which one or more of the premises is suppressed.

This is the usual form of an argument. Perfectly formal syllogisms are very seldom met with. But in order that an argument which has not the form of a perfect syllogism may be valid it must be capable of being put into the form of regular syllogisms. It should also be observed that, though one or more premises may be suppressed, no *term* must be wanting.

The different kinds of abridged syllogisms which we will consider are:

1. *The Enthymeme.*
2. *The Epichirema.*
3. *The Sorites.*

1. ENTHYMEME.—*An enthymeme is an abridged simple syllogism in which one or both of the premises is suppressed.*

The ethymeme is of two kinds.

a) Either the major or the minor premise is suppressed. For instance,

The square is a parallelogram.
∴ The opposite angles of a square are equal.

All men are mortal.
∴ Napoleon is mortal.

In the first example the major premise, and in the second the minor premise is suppressed.

b) Both premises are suppressed, the middle term being included in the conclusion. For instance,

The square, being a parallelogram, has the opposite sides equal.

The enthymeme has very often the form of a sentence consisting of two propositions, united by the conjunction *because.* Thus, *Napoleon is mortal because he is a man* is really an enthymeme. It can easily be put into the form given above.

2. EPICHIREMA.—*An epichirema is an abridged compound syllogism is which one or more of the premises are enthymemes.*

For instance,
    All minerals, being material bodies, have weight.
    Gold, being a metal, is a mineral.
    ∴ Gold has weight.

This may be put into the regular syllogistic form as follows:

        All material bodies have weight.
1.   All minerals are material bodies.
    ∴ All minerals have weight.

        All metals are minerals.
2.   Gold is a metal.
    ∴ Gold is a mineral.

        All minerals have weight.
3.   Gold is a mineral.
    ∴ Gold has weight.

3. SORITES.—*A sorites or chain-argument is an abridged poly-syllogism consisting of three or more simple premises.*

There are two kinds of sorites, the *Aristotelian* and the *Goclenian*, the former having been invented by *Aristotle* and the latter by *Goclenius*. These two kinds of sorites may be stated in the following way:

Aristotelian.

All A are B.
All B are C.
All C are D.
All D are E.
∴ All A are E.

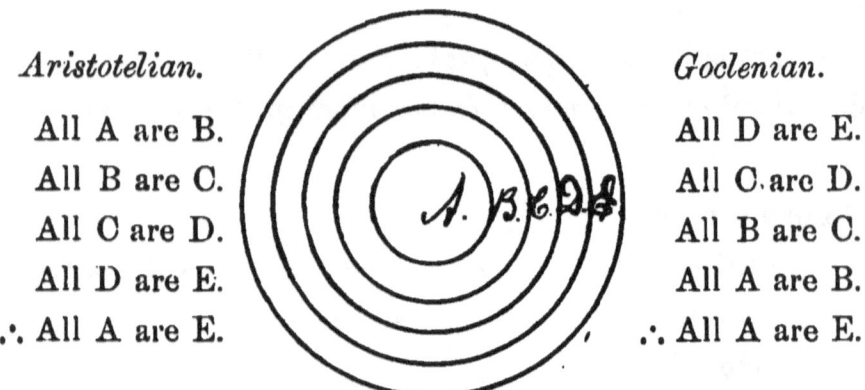

Goclenian.

All D are E.
All C are D.
All B are C.
All A are B.
∴ All A are E.

In the *Aristotelian* sorites the predicate of one premise becomes the subject of the next, and the conclusion has for its subject the subject of the first premise, and for its predicate the predicate of the last premise.

In the *Goclenian* sorites the order is reversed. The subject of one premise becomes the predicate of the next, and the conclusion has for its subject the subject of the last premise and for its predicate the predicate of the first premise.

In the *Aristotelian* sorites we go from the term of least extent to the term of greatest extent, and in the *Goclenian* sorites from the term of greatest extent to the term of least extent. Therefore the former is also called an *ascending* sorites and the latter a *descending* sorites.

EXAMPLES.

Aristotelian sorites.

All flies are insects.
All insects are invertebrates.
All invertebrates are animals.
All animals are organic beings.
∴ All flies are organic beings.

Goclenian sorites.

All animals are organic beings.
All invertebrates are animals.
All insects are invertebrates.
All flies are insects.
∴ All flies are organic beings.

In regard to the quality and quantity of the premises, it should be observed that in the Aristotelian sorites the only premise that may be *particular* is the *first*, and the only one that may be *negative* is the *last*. The Aristotelian sorites given above may be put into the syllogistic form as follows:

1.	2.	3.
B is C	C is D	D is E
A is B	A is C	A is D
∴ A is C	∴ A is D	∴ A is E

The simple syllogisms of which the sorites is composed are all in the first figure, and in this figure the major premise must be universal and the minor premise affirmative. Hence the first premise of the sorites, being the only *minor* premise expressed, is the only one that may be particular.

Again, the last premise of the sorites is the only one that may be negative. For if any other be negative, the conclusion of the corresponding simple syllogism would be negative, and as this conclusion is to be used as the minor premise of the next syllogism, we would have a syllogism in the first figure having a negative minor premise, which is contrary to the rule.

CHAPTER V.

FALLACIES.

1. Definition of Fallacy.

A fallacy is an argument which at first sight appears to be valid, but in reality violates the rules of the syllogism.

2. Classification of Fallacies.

Fallacies are usually divided into two classes: *logical fallacies* and *material fallacies*.

a) A logical fallacy is one in which the premises are insufficient or where the conclusion does not follow from the premises.

b) A material fallacy is one in which the premises are sufficient for the conclusion, but in which either the truth of the premises remains to be proved or the conclusion is irrelevant to the point that is to be demonstrated.

A material fallacy is not a fallacy in the *form*, but in the *subject-matter*. To decide whether the premises are true or not, is something that logic cannot do. The subject of material fallacies is therefore one with which logic is only indirectly concerned.

3. Logical Fallacies.

Logical fallacies may be divided into *purely logical* and *semi-logical*.

Of *purely logical* fallacies, including all the distinct violations of the syllogism, the following may be mentioned:

1. **Fallacy of Four Terms.**
2. **Fallacy of Undistributed Middle.**
3. **Fallacy of Illicit Process of either Major or Minor Term.**
4. **Fallacy of Negative Premises.**
5. **Fallacy of Particular Premises.**

All these fallacies are explained in the chapter treating of syllogisms. We will only give the following examples:

1. A is B.
   C is D.
   ∴ D is A.

Here we have no middle term or medium of comparison between A and D. Hence in order to compare A and D two syllogisms are required, one for comparing A and C with B and the other for comparing A and C with D.

2. All birds are vertebrates.
   All fishes are vertebrates.
   ∴ All fishes are birds.

3. All insects are animals.
   No dogs are insects.
   ∴ No dogs are animals.

No birds are quadrupeds.
4. No horses are birds.
∴ No horses are quadrupeds.

Some flowers are blue.
5. Some flowers are red.
∴ Some red things are blue.

Of *semi-logical* fallacies the more common are:

1. **Fallacy of Equivocation.**
2. **Fallacy of Composition.**
3. **Fallacy of Division.**
4. **Fallacy of Accident.**
5. **Converse Fallacy of Accident.**
6. **Fallacy of Many Questions.**
7. **Fallacy of Amphibology.**
8. **Fallacy of Positive and Negative Intention.**

1. FALLACY OF EQUIVOCATION.—*This fallacy consists in using a term in two different senses.*

In most cases it is the middle term that is used in two different significations in the premises. In such a case the fallacy is usually called a *fallacy of ambiguous middle.* The fallacy of equivocation is, in reality, a fallacy of four terms, as is easily seen by substituting some other expression for the ambiguous term in each premise. For instance,

> No designing person ought to be trusted.
> Engravers are designers.
> ∴ Engravers ought not to be trusted.

> A ball is a round body.
> He attended the ball.
> ∴ He attended a round body.

2. FALLACY OF COMPOSITION.—*This fallacy consists in using the middle term distributively in the major premise and collectively in the minor premise.*

For instance,

> Five and three are two numbers.
> Eight is five and three.
> ∴ Eight is two numbers.

3. FALLACY OF DIVISION.—*This fallacy consists in using the middle term collectively in the major premise and distributively in the minor premise.*

For instance,

> Eight is one number.
> Five and three are eight.
> ∴ Five and three are one number.

> All the apples in the garden are worth one hundred dollars.
> This is one of the apples in the garden.
> ∴ This apple is worth one hundred dollars.

4. FALLACY OF ACCIDENT.—*This fallacy consists in asserting of something described by some accidental peculiarity what is true only of its substance.*

For instance,
>What you bought yesterday you eat to-day.
>You bought raw meat yesterday.
>∴ You eat raw meat to-day.

We do not buy meat because it is raw, but because it is meat. That the meat is raw is only an accidental property.

5. CONVERSE FALLACY OF ACCIDENT.—*This fallacy consists in arguing from a special case to a general one.*

For instance,
>Alcohol acts as a poison when used in excess.
>∴ Alcohol is always a poison.

6. FALLACY OF MANY QUESTIONS.—*This fallacy consists in combining two or more questions into one to which a single answer cannot be given.*

Thus, if a man who has never used tobacco is asked *If he has given up smoking*, he can neither answer the question affirmatively nor negatively. This question would namely imply that he did smoke. This fallacy arises from the fact that though only one question is expressed, two or more questions are implied.

7. Fallacy of Amphibology.—*This fallacy consists in ambiguity in the grammatical structure of a sentence, by which it may have two or more different meanings.*

Thus a word may be used so as to leave it ambiguous whether it is subject or predicate, or the reference of a pronoun or an adverb may be ambiguous. For instance,

> He likes me better than *you*.

> We also get salt from the ocean, *which* is very useful to man.

> He promised his father to help *his* friends.

8. Fallacy of Positive and Negative Intention.—*This fallacy consists in using certain negative words, as no and nothing, in two different senses.*

For instance,
> No cat has two tails.
> Every cat has one tail more than no cat.
> ∴ Every cat has three tails.

> Nothing is better than happiness.
> Bread is better than nothing.
> ∴ Bread is better than happiness.

4. Material Fallacies.

Of material fallacies the more common are:

1. Begging the Question (Petitio principii).

2. **Fallacy of False Cause** (Non causa pro causa).
3. **Fallacy of Irrelevant Conclusion** (Ignoratio elenchi).

1. BEGGING THE QUESTION.—*This fallacy consists in using as a premise either the conclusion itself or some consequence of the conclusion which is to be established.*

Another name for this kind of fallacy is *arguing in a circle* (circulus in demonstrando). Thus we argue in a circle if we try to prove the existence of God in the following way:

> The Scriptures must be true, as they are the word of God.
>
> The Scriptures declare that God exists.
>
> ∴ God exists.

Here we prove that God exists from the truth of the Scriptures and prove the truth of the Scriptures from the fact that they are the word of God, which evidently implies that we take for granted what is to be proved, namely, that God exists.

2. FALLACY OF FALSE CAUSE.—*This fallacy consists in assigning as a cause something that, in reality, has nothing to do with the conclusion.*

If one event occurs shortly before another event or they occur at the same time, and if we take the mere conjunction of the two events as a satisfactory proof that one is the cause of the other, we commit a fallacy of false cause. Two events may be simultaneous without having the least relation.

3. FALLACY OF IRRELEVANT CONCLUSION.—*This fallacy consists in arriving at a conclusion different from the one that is to be established.*

Suppose we had to prove that all the angles of a triangle are together equal to two right angles, and we only proved that they cannot be less than two right angles. That would be a fallacy of irrelevant conclusion, for the proposition would not be proved before we had also proved that the angles cannot be more than two right angles.

The fallacy of irrelevant conclusion is one of the most common of the material fallacies, and is known under various names. Of the more common forms of this kind of fallacy the following two may be mentioned:

a) *Argumentum ad hominem*, which consists in making an appeal to the vanity or prejudice of our opponent so as to make him blind to the unreasonableness of the argument.

b) *Argumentum ad populum*, which differs from the former fallacy only in being addressed to a body of people instead of one individual.

5. Paralogisms and Sophisms.

Fallacies may also be divided into *paralogisms* and *sophisms*.

1. A *paralogism* is an undesigned fallacy, the person that commits it being unconscious of the falsity of his argument.

2. A *sophism* is a fallacy which is consciously used to deceive.

CHAPTER VI.

METHOD.

1. Science.

1. DEFINITION OF SCIENCE.—*Science is classified knowledge.*

A person may have learned a good many facts about a certain group of objects or phenomena, but in order that his knowledge may be entitled to the name of scientific knowledge, the facts must be arranged according to certain principles and the relation between them clearly understood.

Scientific knowledge does not differ in kind from common knowledge, as the powers used in acquiring knowledge, whether it be common or scientific, must obviously be the same. They differ only in degree of accuracy.

2. REQUISITES OF A SCIENCE.—The requisites of a science are:

a) *All statements made must be true.*

b) *A science should be as general as possible; i. e., the process of generalization should be carried as far as possible.*

c) *In every science there should be a certain order, and a necessary connection between the various elements of the science.*

d) *The number of facts ascertained should be as great as possible.*

3. Axioms.—*An axiom is a self-evident and intuitively true proposition.*

The truth of an axiom cannot and need not be demonstrated by any simpler propositions.

The ultimate principles of all deductive sciences are axioms, which form the basis on which all the demonstrations of those sciences are founded. As examples of axioms we may mention the following two:

The whole is greater than its parts.

Things that are equal to the same thing are equal to each other.

2. Deduction and Induction.

1. Definition of Method.—*Method is a certain mode of procedure for arriving at a certain result.*

Method must be used in all sciences, though the kind of method which is to be used will be different for different sciences.

The methods used in science may be classified under the two heads, *deduction* and *induction*.

2. Deduction.—*Deduction is the process of deriving a particular truth from a general truth.*

In the deductive method we proceed from the general to the particulars which are embraced in it.

For instance,

> All insects are animals.
> All butterflies are insects.
> ∴ All butterflies are animals.

Here we first state a general truth, something that is true about all insects, namely, that they are animals. Then we proceed to analyze this general truth into the particulars it embraces, and finally we reach a conclusion concerning one of the particulars, namely, butterflies.

The deductive method is also called the *analytic method*.

3. Induction.—*Induction is the process of deriving general truths from particular truths.*

In the inductive method we proceed from the observation of particular truths or facts to the establishment of general laws. As an example of inductive reasoning we give the following:

> By observations we know that Mercury, Venus, the Earth, Mars, Jupiter, Saturn, Uranus and Neptune move around the sun in elliptic orbits.
>
> Hence all the planets move around the sun in elliptic orbits.

The inductive method is also called the *synthetical method*.

Induction is of two kinds: *Perfect induction* and *imperfect induction*.

a) The induction is perfect when all the particular cases have been examined.

For instance,

> Mercury, Venus, the Earth, Mars, etc., move in elliptic orbits around the sun.
> Hence all the known planets move around the sun in elliptic orbits.

In the conclusion we affirm something only of the particular cases that have been examined. We do not say that *all* planets move in elliptic orbits around the sun, but only all the *known* planets. The conclusion must therefore be certain.

Perfect induction always leads to a necessary and certain conclusion.

b) The induction is imperfect when we have examined only some of the particular cases and from them infer a general law.

In the first example given above we assert of *all* planets something that has been found to be true of all the *known* planets. Hence we infer that if some new planet would be discovered it would most likely move in an elliptic orbit around the sun like the planets that are now known. This conclusion is very probable, but not certain.

Imperfect induction can never lead to a certain and necessary conclusion, but only to a probable conclusion.

3. Definition.

1. DEFINITION DEFINED.—*To define a thing is to give those attributes by which it differs from all other things, and the process is called logical definition.*

To define something means to state what it is, or to distinguish it from all other things. It is not necessary, however, to enumerate all the attributes belonging to the thing which is to be defined, but only the *essential attributes*. The essential attributes are the *genus* and the *differentia*.

a) By the genus is meant the next higher genus of which the thing to be defined is a species.

b) By the differentia is meant those specific characters by which the thing to be defined differs from all other species of the same genus.

Definition thus consists in giving the genus and the differentia of the thing to be defined. A definition has the form of a categorical judgment, of which the subject is the thing to be defined, and the predicate the genus and the differentia.

Suppose we want to define an *equilateral triangle*. An equilateral triangle is a species of the genus triangle, and differs from all other triangles in having the three sides equal. Hence the genus is *triangle* and the differentia *having the three sides equal.* The definition of an equilateral triangle will then be:

An equilateral triangle is a *triangle* [genus] having its sides equal. [differentia]

It should be observed that it is essential for a logical definition that the genus should be the next higher genus. Hence the following definition is not correct:

An equilateral triangle is a plane figure having its sides equal.

Plane figure is not the next higher genus.

We will give two more examples of definitions, viz.:

Man is a rational [differentia] animal [genus].

A parallelogram is a quadrilateral [genus] whose opposite sides are parallel. [differentia]

2. RULES FOR DEFINITION.—In definition the following rules should be observed:

I. *The definition should be adequate, i. e., neither too wide nor too narrow.*

a) The definition is too *wide* if the predicate has greater extent than the subject, *i. e.*, if it includes other things besides those that are to be defined.

For instance,

A bird is an animal that has a backbone.

This definition is too wide because also fishes, reptiles and mammals have a backbone.

Man is a rational being.

This is also too wide, because *rational* being also includes God.

b) The definition is too *narrow* if the subject has a greater extent than the predicate, *i. e.*, if it excludes some of the things that are to be defined.

For instance,

A triangle is a figure having three equal sides.

This definition is too narrow, because all isosceles and scalene triangles are excluded.

A bird is a feathered animal that sings.

This is also too narrow. Some birds do not sing.

The test of an adequate definition is that it may be both simply converted and converted by contraposition. If the definition is too wide, it cannot be simply converted. If it is too narrow, it cannot be converted by contraposition.

II. *The definition should not contain the term which is to be defined.*

The violation of this rule is called *defining in a circle.* We thus define in a circle if we define *law* as *a lawful command*, because we use in the definition the word we want to define. As another example let us take

Life is the sum of the vital functions.

Here we use the term *vital*, which is really a synonym of the term to be defined, and which only can be explained by the term *life*.

III. *The definition should be affirmative.*

The definition should state what a thing is, and not what it is not. Hence the following definitions are unsatisfactory:

A straight line is a line no portion of which is curved.

A regular polygon is one that is not irregular.

Light is the absence of darkness.

IV. *In definition we should not give any superfluous or accidental attributes.*

For instance,

A pentagon is a polygon having five sides and five angles.

This definition is incorrect, as the latter attribute is superfluous.

A parallelogram is a quadrilateral having the opposite sides parallel and having the opposite sides and angles equal.

Here two attributes are given that follow from the parallelism of the sides, and which therefore are superfluous.

>A horse is a four-legged animal with a tail and a mane.

Here accidental attributes are used.

3. NOMINAL AND REAL DEFINITIONS.—Definitions are divided into *nominal* and *real.*

a) A nominal definition is one which explains the meaning of the term which is used as the name of the thing.

For instance,
>A phonograph is an instrument for registering and reproducing sound.
>
>A telephone is an instrument for conveying sound to a great distance.

b) A real definition is one which defines the thing itself.

Thus a real definition of *phonograph* would be a treatise on the construction and use of that instrument.

In all scientific investigations it is the aim to obtain real definitions, but for many practical purposes nominal definitions will be sufficient.

4. DESCRIPTION.—*By description is meant an enumeration of all the properties of a thing.*

A description of an elephant, for instance, would thus consist in the enumeration of all the properties belonging to elephants. In definition we give only the essential attributes of a thing. In description, again, we may use not only essential, but also accidental attributes. The natural-history sciences furnish good examples of descriptions.

4. Division.

1. DIVISION DEFINED.—*By logical division is meant the process of dividing a genus into its species according to a certain principle of division.*

For instance,

Triangles may be divided into
$\begin{cases} \text{right-angled} \\ \text{acute-angled} \\ \text{obtuse-angled.} \end{cases}$

Here the genus triangle is separated into its three species, and the basis or principle of division, commonly called the *fundamentum divisionis*, is the size of the angles.

Polygons may be divided into
$\begin{cases} \text{triangles} \\ \text{quadrilaterals} \\ \text{pentagons} \\ \text{hexagons} \\ \text{etc.} \end{cases}$

Here the principle of division is the number of sides.

2. DICHOTOMY.—If a genus is divided into two species each of which is the contradictory of the other, the division is commonly called *dichotomy.*

For instance,

Animals may be divided into
$\begin{cases} \text{vertebrates} \\ \text{not-vertebrates.} \end{cases}$

Polygons may be divided into
$\begin{cases} \text{triangles} \\ \text{not-triangles.} \end{cases}$

Although, from a logical point of view, dichotomy is a perfect division, it is for most practical purposes not very convenient.

3. RULES FOR DIVISION.—In logical division the following rules should be observed:

I. *In division there should be only one principle of division.*

Hence the following divisions are not correct:

Books are divided into { English, French, German, Quarto, Octavo, etc. }

The first division is according to language and the second according to size.

Triangles are divided into { isosceles, equilateral, right-angled, acute-angled. }

The first division is according to the relative length of the sides and the second according to the size of the angles.

Such a division is generally called a *cross-division*.

II. *The principle of division should be an actual attribute of the genus which is to be divided.*

III. *In division the members should exclude each other, and they should all be co-ordinate or of the same rank.*

For instance,

Polygons may be divided into
$\begin{cases} \text{triangles} \\ \text{quadrilaterals} \\ \text{parallelograms} \\ \text{polygons having more} \\ \quad \text{than four sides.} \end{cases}$

Parallelograms are included in quadrilaterals, and consequently the members do not exclude each other.

IV. *The division should be complete, i. e., the sum of the species should be equal to the genus.*

Hence no species must be left out. For instance,

Vertebrates are divided into $\begin{cases} \text{mammals} \\ \text{birds} \\ \text{fishes.} \end{cases}$

Here *reptiles* and *batrachians* are left out.

Triangles are divided into $\begin{cases} \text{acute-angled} \\ \text{right-angled.} \end{cases}$

Here *obtuse-angled* triangles are left out.

V. *In division we should proceed from proximate genera to proximate species.*

We should not proceed from a high genus to a low species, but from the genus to the next lower species.

In the following division this rule is violated:

Vertebrates are divided into
{
horses
dogs
eagles
lions
etc.
}

A logical division of vertebrates would be into
{
mammals
birds
fishes
batrachians
reptiles.
}

Each of these species may further be divided and subdivided until we reach the lowest species.

4. PARTITION.—*By partition is meant the separation in thought of the physical parts of which an individual object is composed.*

For instance,

Water is composed of oxygen and hydrogen.

A plant may be divided into root, stem, leaves, etc.

This mode of separating an object into its constituent parts is something with which logic is not directly concerned, and should not be confounded with logical division.

5. Demonstration.

1. DEMONSTRATION DEFINED.—*Demonstration is an act of reasoning by which the truth of a proposition is established as a consequence of other truths.*

In every demonstration we notice:

a) The proposition that is to be proved.

b) The premises or grounds of proof.

c) The necessary connection between the different parts of the demonstration.

The premises are either definitions, axioms, or previously established propositions.

2. RULES FOR DEMONSTRATION.—For demonstration we have the following rules:

I. *No proposition must be used as a premise which is not known to be true.*

II. *The proposition which is to be proved must not be used as a premise.*

III. *No proposition whose truth depends on the truth of the proposition which is to be proved must be used as a premise.*

IV. *There must be no leaps in the demonstration.*

V. *We must not prove another proposition instead of the one that is to be established.*

For violations of the rules given above see Chapter V (*Fallacies*).

3. Classification of Demonstrations. — Demonstrations are divided —

    I. Into *direct* and *indirect.*
    II. Into *deductive* and *inductive.*
    III. Into *a priori* and *a posteriori.*

I. *a) A direct demonstration is one in which the truth of a proposition is immediately deduced from certain other truths that have already been established.*

b) An indirect demonstration is one in which the truth of a proposition is established by proving the absurdity of its contradictory.

In a direct demonstration we give the reasons why the conclusion must be true. In an indirect demonstration we give the reasons why it cannot be false. In an indirect demonstration we proceed in the following manner. We make a supposition contrary to the conclusion which is to be proved. From this supposition we deduce a series of conclusions until we arrive at a conclusion which is contrary to some known truth. Then by modus tollens we conclude from the falsity of the consequent to the falsity of the antecedent; that is, we conclude that the supposition made must be false, as it leads to an absurd conclusion. And since this supposition is false, its contradictory, or the conclusion which is to be established, must be true; because of two contradictories one must be true and the other false.

We give the following example of an indirect demonstration:

Two straight lines perpendicular to the same straight line are parallel.

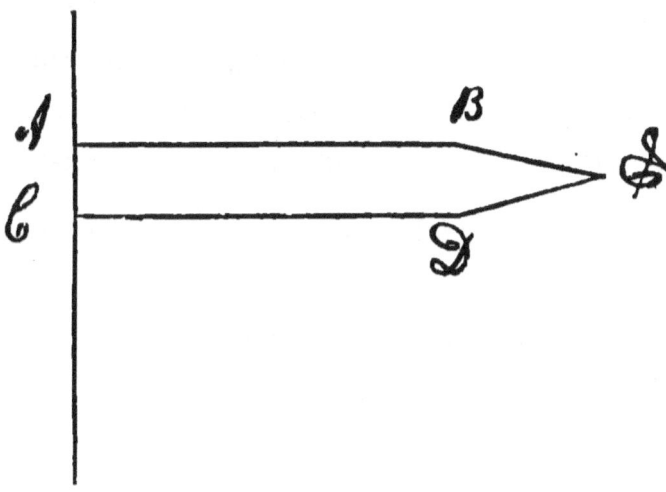

Let the two straight lines AB and CD be both perpendicular to AC; then AB is parallel to CD.

For suppose that AB is not parallel to CD. Then the two lines AB and CD must meet at some point if they be produced. Let them meet at the point E. Then there will be two perpendiculars, EA and EC, let fall from the same point on the same straight line, which is absurd. Therefore the two lines, AB and CD, cannot meet if they be produced ever so far. Hence the two lines are parallel.

II. *a) A deductive demonstration is one in which we proceed from the whole to the parts.*

b) An inductive demonstration is one in which we proceed from the parts to the whole.

In a deductive demonstration we prove that something holds true of the whole, and then conclude that it must hold true of every part or individual case of the whole. In an inductive demonstration, again, we prove that something holds true of all the parts or individual cases and then conclude that it must hold true of the whole.

We will give the following example of an inductive demonstration:

An angle inscribed in a segment is measured by half the arc included between its sides.

This proposition admits of three cases:

1st. Let the centre of the circle be on one of the sides of the angle.

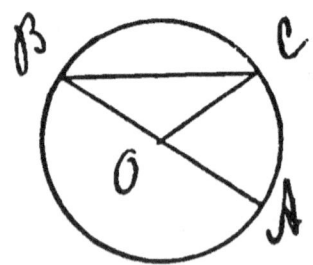

Draw the radius OC. Because OC is equal to OB, the angle OBC is equal to the angle OCB; therefore the angles OBC and OCB are together double the angle OBC. The angle AOC is equal to the sum of the angles OBC and OCB. Hence the angle AOC is double the angle OBC. But the angle AOC is measured by the arc AC. Hence the angle ABC is measured by half the arc AC.

2d. Let the centre of the circle be within the angle.

Draw the diameter BD. By the first case we know that the angle ABD is measured by half the arc AD and the angle DBC by half the arc DC. Therefore the angle ABC is measured by half the sum of the arcs AD and DC, *i. e.*, half the arc AC.

3d. Let the centre be w.thout the angle.

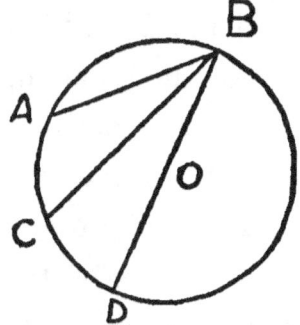

Draw the diameter BD. By the first case we know that the angle ABD is measured by half the arc AD, and the angle CBD by half the arc CD. Therefore the angle ABC is measured by half the difference of the arcs AD and CD, *i. e.*, half the arc AC.

Hence the proposition is true for all possible cases, and therefore it must be true for any angle inscribed in a segment.

III. *a) A demonstration a priori is one in which the premises are given by intuition.*

b) A demonstration a posteriori is one in which the premises are given by experience.

In mathematics, for instance, all the relations between quantities are established by a chain of reasoning which ultimately depends on certain a priori or intuitive principles, namely the ideas of space and number. In the natural sciences, again, the arguments are mainly a posteriori, as the premises are given by experience.

6. Analogy.

Reasoning by analogy is a process by which we infer that if two or more objects are similar in certain respects, they will also be similar in other respects.

Reasoning by analogy gives only a probable conclusion. The degree of probability depends on the number of observed resemblances and the importance of the points in which the objects agree. Hence in order that reasoning by analogy should be of any value, the attributes that are similar should be as many as possible and should not be accidental. If it can be shown that one or more of the essential attributes of the first object is incompatible with some essential attribute of the second object, the argument is invalid.

For instance,

> By observing the similarity between lightning and electricity in many respects, Franklin was, by analogy, led to the conclusion that they were identical.

> The earth and the planet Mars resemble each other in many respects.
> Hence Mars is probably inhabited.

7. Hypothesis.

A hypothesis is a supposition made to account for a certain group of phenomena.

The probability of a hypothesis depends on the number of facts or phenomena that may be explained by it. The greater the number of phenomena it will explain, the more we are justified to believe the hypothesis to be right.

As examples of hypotheses we may mention Laplace's Nebular Hypothesis to explain the formation of the solar system, and the Copernican theory of the solar system.

8. Classification of Sciences.

From a formal point of view the sciences are usually divided into *empirical* and *rational.*

The difference between the empirical and the rational sciences is given in the following schedule:

Empirical
- *Data:* facts.
- *Aim:* the establishment of general laws.
- *Method:* mainly inductive.

Rational
- *Data:* universal principles.
- *Aim:* the establishment of particular truths.
- *Method:* mainly deductive.

Botany, zoology, chemistry and geology are examples of empirical sciences. Mathematics is an example of a rational science.

EXERCISES.

CHAPTER II.

CLASSIFICATION OF CONCEPTS.

1. For each one of the following concepts state whether it is positive or negative, absolute or relative, concrete or abstract:

Book	Man	Daughter
Father	House	Metal
Weight	Darkness	Independence
Holiness	Logic	Whiteness
Unnatural	Light	Son
Air	Resemblance	Animal
Oblique-angled	Curved	Straight
Being	Reason	Rational
Figure	Triangle	God

EXTENT AND CONTENT.

1. In each one of the following pairs of concepts state which concept has the greater extent, and which has the greater content:

$1\begin{cases} \text{Dog} \\ \text{Animal} \end{cases}$
$2\begin{cases} \text{Plant} \\ \text{Tree} \end{cases}$
$3\begin{cases} \text{Man} \\ \text{Being} \end{cases}$

EXERCISES. 93

4 { Heavenly body / Planet   5 { Element / Metal   6 { Eagle / Bird

7 { Equitable triangle / Equiangular triangle   8 { Fish / Vertebrate   9 { Rock / Granite

10 { Fly / Insect   11 { Book / Dictionary   12 { House / Brick house

2. Arrange the following terms in several series in such a manner that the first term of each series shall have the greatest extent and the last term the least extent.

Salmon	Plant	Europeans
Polygon	Fish	Square
Man	Animal	Figure
Apple tree	Vertebrate	Rational being
Plane figure	Quadrilateral	Phœnogam

RELATION OF CONCEPTS.

1. State the relation between the concepts of each of the following groups:

1 { Plant / Organic being   2 { Polygon / Figure   3 { Flies / Bees

4 { Square / Parallelogram   5 { Man / American   6 { European / Italian

7 { Metal / Not-metal   8 { Gold / Iron   9 { Salmon / Fish

10 { Bird / Reptile   11 { Straight / Not-straight   12 { Eagle / Sparrow

CHAPTER III.

CLASSIFICATION OF JUDGMENTS.

State the logical character as to quality, quantity, relation and modality of each of the following judgments:

> All triangles are figures.
> If he is honest, he should speak the truth.
> Triangles are divided into right-angled and oblique-angled.
> The table is black.
> If rain has fallen, the ground is wet.
> Napoleon was a great man.
> No triangles are squares.
> Some angles are obtuse.
> Some horses are not black.
> His character is either good or bad.
> Iron is an element.
> Some men are good.
> God is omniscient.
> Some men are not kings.
> This horse is not black.
> Some triangles are equilateral.
> No planets are self-luminous.
> Some of our muscles are involuntary.
> New York is a city.
> Horses are vertebrate animals.

IMMEDIATE INFERENCE.

1. Which of the four judgments A, I, E and O are true or false when

    1. A is true
    2. A is false
    3. I is true
    4. I is false
    5. E is true
    6. E is false
    7. O is true
    8. O is false

2. Convert the following judgments:

    All vertebrates are animals.
    Some poisonous things are plants.
    No men are angels.
    Man is mortal.
    Some persons are wise.
    Some quadrupeds are not horses.
    Some birds are eagles.
    No plants are animals.
    All triangles have three sides.
    No triangles are quadrilaterals.

3. If the judgment

    Some triangles are not figures

    is false, how could you prove the truth of the judgment

    Some triangles are figures?

4. How can you conclude from the falsity of the judgment

    No animals living in water are fishes

    to the truth of

    Some fishes live in water?

5. How can you conclude from the truth of the judgment
    No insects are vertebrates
  to the falsity of
    Some vertebrates are insects?

6. How can you prove the falsity of the judgment
    No not-triangles are figures
  from the truth of
    Some figures are not triangles?

CHAPTER IV.

SIMPLE SYLLOGISMS.

Construct syllogisms from the terms given in each of the following moods:

1st Figure.

$Barbara \begin{cases} P = \text{animal} \\ M = \text{bird} \\ S = \text{eagle} \end{cases}$
$Celarent \begin{cases} P = \text{irrational} \\ M = \text{man} \\ S = \text{American} \end{cases}$

$Darii \begin{cases} P = \text{mortal} \\ M = \text{man} \\ S = \text{being} \end{cases}$
$Ferio \begin{cases} P = \text{square} \\ M = \text{triangle} \\ S = \text{equilateral figure} \end{cases}$

2d Figure.

$Cesare \begin{cases} P = \text{animal} \\ M = \text{plant} \\ S = \text{grass} \end{cases}$
$Camestres \begin{cases} P = \text{insect} \\ M = \text{animal} \\ S = \text{rock} \end{cases}$

Festino $\begin{cases} \text{P=trapezium} \\ \text{M=parallelogram} \\ \text{S=quadrilateral} \end{cases}$ *Baroko* $\begin{cases} \text{P=fixed star} \\ \text{M=selfluminous} \\ \text{S=heavenly body} \end{cases}$

3d Figure.

Darapti $\begin{cases} \text{P=man} \\ \text{M=American} \\ \text{S=rational} \end{cases}$ *Disamis* $\begin{cases} \text{P=hawk} \\ \text{M=vertebrate} \\ \text{S=animal} \end{cases}$

Datisi $\begin{cases} \text{P=mortal} \\ \text{M=man} \\ \text{S=black} \end{cases}$ *Felapton* $\begin{cases} \text{P=herb} \\ \text{M=tree} \\ \text{S=plant} \end{cases}$

Bokardo $\begin{cases} \text{P=having feet} \\ \text{M=reptile} \\ \text{S=animal} \end{cases}$ *Ferison* $\begin{cases} \text{P=triangle} \\ \text{M=pentagon} \\ \text{S=equilateral} \end{cases}$

4th Figure.

Bramantip $\begin{cases} \text{P=granite} \\ \text{M=rock} \\ \text{S=inorganic} \end{cases}$ *Camenes* $\begin{cases} \text{P=European} \\ \text{M=man} \\ \text{S=plant} \end{cases}$

Dimaris $\begin{cases} \text{P=yellow} \\ \text{M=butterfly} \\ \text{S=insect} \end{cases}$ *Fesapo* $\begin{cases} \text{P=cat} \\ \text{M=dog} \\ \text{S=animal} \end{cases}$

Fresison $\begin{cases} \text{P=square} \\ \text{M=hexagons} \\ \text{S=equilateral} \end{cases}$

COMPOUND SYLLOGISMS.

1. From the following terms construct compound syllogisms, epichiremata, and sorites:

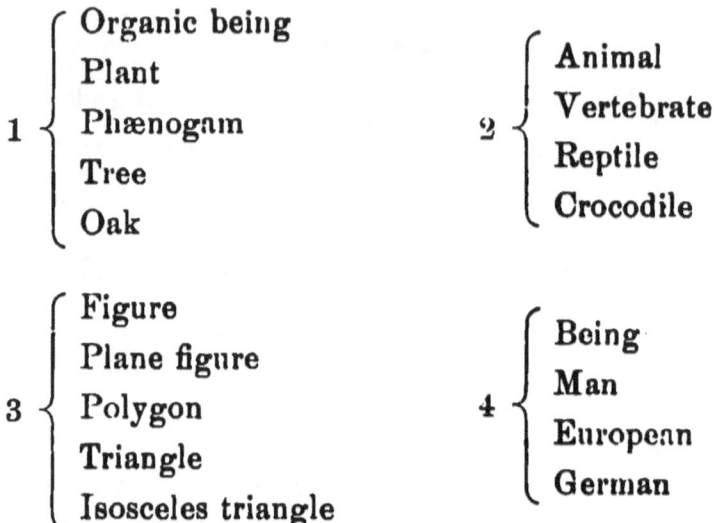

2. Construct enthymes by taking any three consecutive terms of those given above.

CHAPTER V.

FALLACIES.

Point out the fallacies in the following arguments:

1. Some plants are trees.
   Some plants are grasses.
   ∴ Some grasses are trees.

2. Red is a color.
Blue is a color.
∴ Blue is red.

3. All men are rational beings.
All men are animals.
∴ All animals are rational beings.

4. All men are organic beings.
No dogs are men.
∴ No dogs are organic beings.

5. All moral beings are accountable.
No brute is a moral being.
∴ No brute is accountable.

6. Design implies a designer.
The universe abounds in design.
∴ God exists.

7. A stone is a body.
An animal is a body.
Man is an animal.
∴ Man is a stone.

8. Nothing is better than wisdom.
A dime is better than nothing.
∴ A dime is better than wisdom.

9. Metals are elements.
   Iron is a metal.
   ∴ Iron is an element.

10. If this medicine is of any value, those who take it will improve in health.
    I have taken it, and have improved in health.
    ∴ This medicine is of value.

11. Dickens's Oliver Twist is one of the books in the book-store of my friend.
    I have bought Dickens's Oliver Twist.
    ∴ I have bought one of the books in my friend's book-store.

12. His books are worth one hundred dollars.
    Shakespeare is one of his books.
    ∴ Shakespeare is worth one hundred dollars.

13. The people of the city are suffering from the yellow fever.
    You are one of the people of the city.
    ∴ You are suffering from the yellow fever.

14. Light is contrary to darkness.
    Feathers are light.
    ∴ Feathers are contrary to darkness.

CHAPTER VI.

DEFINITION.

1. Define the following terms, and point out the *genus* and the *differentia* in each definition:

Element	Capital	Dictionary
Nonagon	Vertebrate	Genus
Logic	Animal	Species
Science	Man	Hypothesis
Syllogism	Rhombus	Circle
Deduction	Plant	Straight line
Induction	Parallelogram	Judgment

2. What rules do the following definitions violate?

   1. A straight line is one no portion of which is curved.
   2. A rectangle is a figure having four right angles.
   3. A trapezium is a quadrilateral having the opposite sides parallel.
   4. A hexagon is a figure having six equal sides.
   5. A mammal is an animal that does not reproduce its species by laying eggs.
   6. A square is a four-sided figure with equal sides.
   7. Evil is that which is not good.

DIVISION.

In what are the following divisions faulty?

1. Plants are divided into
   - Cryptogams
   - Monopetalous
   - Apetalous
   - Polypetalous

2. Mankind may be divided into
   - Men
   - Women
   - Children

3. Birds are divided into
   - Sea-birds
   - Sparrows
   - Eagles
   - Parrots
   - Gallinaceous birds

4. The faculties of the mind are divided into
   - Perception
   - Imagination
   - Reason

5. Books are divided into
   - Grammars
   - Dictionaries
   - French
   - German
   - Italian

www.ingramcontent.com/pod-product-compliance
Lightning Source LLC
Chambersburg PA
CBHW020148170426
43199CB00010B/937